Welcome!

PAGE A DAY MATH

Thank you for choosing Page A Day Math, a great way to introduce essential math basics and writing numbers. Page A Day Math books help your child develop a solid math foundation through daily step-by-step practice, repetition, and of course, the friendly Math Squad!

How to Use This Book

1. Student traces and solves each problem, completing a page a day, front and back.
2. Parent checks answers and circles incorrect problems.
3. Student corrects errors.
4. Student colors in achievement stars each day when finished!

Have Fun

ISBN – 978-1-947286-04-7

This book belongs to ___________________________

Dear Super Hero Math Student,

You can be a Math Squad Super Hero like Flo, Jo, Bo, Zo, and me! Practice every day and you'll be a math star too!

YOUR FRIEND,
MO

P.S. Check out what my math buddies and I are up to in the Math Squad Monthly at www.PageADayMath.com.

Day 1

PAGE A DAY MATH

Count ⇨ 0 +  = 

Learn ⇨ 0 + 5 = 5

Trace ⇨ 0 + 5 = 5

Copy ⇨ ☐ + ☐ = ☐

Bo says, "Practice and be a math star like me!"

1) 0 + 5 = ☐

2) 8 + 4 = ☐

3) 5 + 0 = ☐

4) 10 + 4 = ☐

5) 4 + 9 = ☐

6) 7 + 4 = ☐

7) 5 + 0 = ☐

8) 4 + 6 = ☐

9) 5 + 4 = ☐

10) 0 + 5 = ☐

Day 1

Wow, you are learning fast. Here are a few more.

11) 5 + 0 = ___

12) 4 + 10 = ___

13) 8 + 3 = ___

14) 0 + 5 = ___

15) 7 + 3 = ___

16) 9 + 4 = ___

17) 4 + 7 = ___

18) 3 + 6 = ___

19) 5 + 0 = ___

20) 3 + 9 = ___

21) 6 + 4 = ___

22) 10 + 3 = ___

23) 4 + 8 = ___

24) 0 + 5 = ___

☆ Color in the stars each day when you finish!

Day 2

Count ⇨ 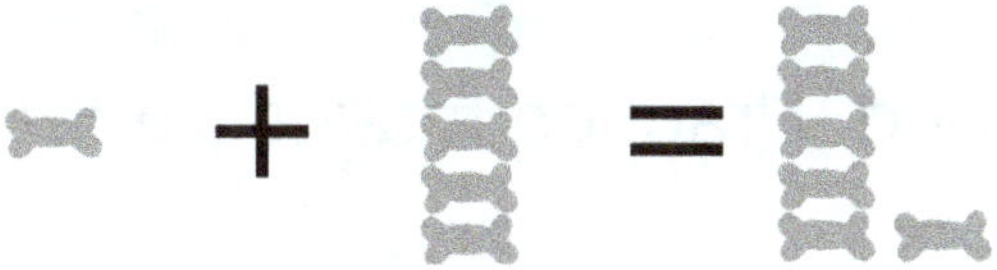

Learn ⇨ 1 + 5 = 6

Trace ⇨ 1 + 5 = 6

Copy ⇨ ☐ + ☐ = ☐

You are on your way to success! Try these!

1) 1 + 5 = ☐

2) 5 + 0 = ☐

3) 9 + 4 = ☐

4) 5 + 1 = ☐

5) 8 + 3 = ☐

6) 0 + 5 = ☐

7) 6 + 4 = ☐

8) 1 + 5 = ☐

9) 4 + 7 = ☐

10) 5 + 1 = ☐

Day 2

PAGE A DAY MATH

You are coming along. Practice makes perfect!

11) 1 + 5 = ☐

12) 6 + 4 = ☐

13) 10 + 4 = ☐

14) 5 + 1 = ☐

15) 4 + 3 = ☐

16) 9 + 4 = ☐

17) 0 + 5 = ☐

18) 6 + 3 = ☐

19) 1 + 5 = ☐

20) 2 + 7 = ☐

21) 5 + 0 = ☐

22) 4 + 5 = ☐

23) 3 + 10 = ☐

24) 4 + 3 = ☐

☆☆ Color in the stars each day when you finish!

Day 3

Count ⇨ 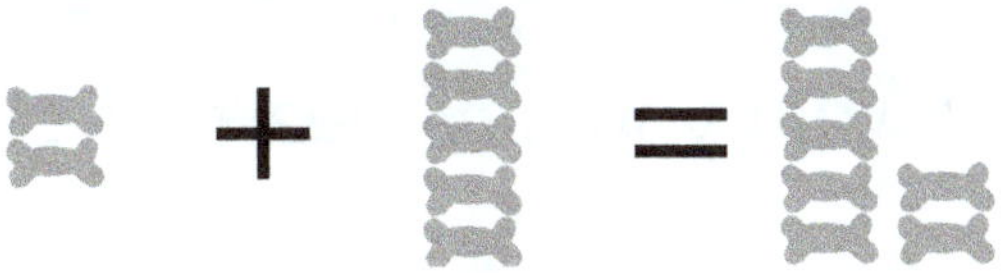

Learn ⇨ 2 + 5 = 7

Trace ⇨ 2 + 5 = 7

Copy ⇨ ☐ + ☐ = ☐

Hurray! Keep going. You've got it.

1) 2 + 5 = ☐

2) 5 + 1 = ☐

3) 5 + 2 = ☐

4) 1 + 5 = ☐

5) 2 + 5 = ☐

6) 5 + 0 = ☐

7) 2 + 5 = ☐

8) 0 + 5 = ☐

9) 1 + 5 = ☐

10) 5 + 2 = ☐

Day 3

PAGE A DAY MATH

Alright! Now review what you have learned so far.

11) 5 + 2 =

12) 0 + 5 =

13) 1 + 5 =

14) 4 + 9 =

15) 2 + 5 =

16) 5 + 1 =

17) 8 + 4 =

18) 5 + 1 =

19) 7 + 3 =

20) 5 + 2 =

21) 0 + 5 =

22) 6 + 3 =

23) 2 + 5 =

24) 4 + 7 =

Color in the stars each day when you finish!

Day 4 Review

PAGE A DAY MATH

Practice makes perfect. That's right!

1) 6 + 4 = ☐

2) 9 + 4 = ☐

3) 4 + 8 = ☐

4) 2 + 7 = ☐

5) 3 + 6 = ☐

6) 4 + 3 = ☐

7) 8 + 2 = ☐

8) 9 + 3 = ☐

9) 3 + 8 = ☐

10) 7 + 3 = ☐

11) 4 + 2 = ☐

12) 5 + 4 = ☐

13) 4 + 7 = ☐

14) 3 + 3 = ☐

Day 4 Review

PAGE A DAY MATH

Keep it up. You are improving with practice! Woof-woof.

15) 9 + 2 = ☐

16) 2 + 6 = ☐

17) 0 + 4 = ☐

18) 2 + 3 = ☐

19) 1 + 2 = ☐

20) 4 + 4 = ☐

21) 5 + 1 = ☐

22) 2 + 7 = ☐

23) 4 + 3 = ☐

24) 6 + 4 = ☐

25) 5 + 2 = ☐

26) 2 + 4 = ☐

27) 5 + 3 = ☐

28) 8 + 2 = ☐

☆☆☆☆ Color in the stars each day when you finish!

Day 5

Count ⇨ 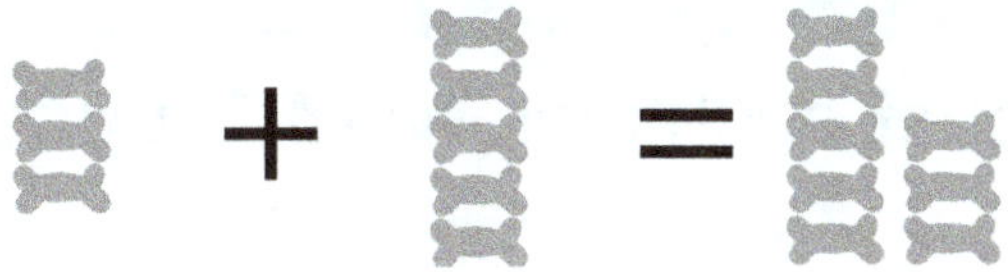

Learn ⇨ 3 + 5 = 8

Trace ⇨ 3 + 5 = 8

Copy ⇨ ☐ + ☐ = ☐

You are doing so well. Keep going. Terrific!

1) 3 + 5 = ☐

2) 5 + 3 = ☐

3) 2 + 5 = ☐

4) 3 + 5 = ☐

5) 5 + 3 = ☐

6) 5 + 2 = ☐

7) 1 + 5 = ☐

8) 5 + 0 = ☐

9) 1 + 5 = ☐

10) 3 + 5 = ☐

Day 5

PAGE A DAY MATH

You are getting better each day. Woof! Yippee!

11) 3 + 5 =

12) 5 + 1 =

13) 2 + 5 =

14) 0 + 5 =

15) 5 + 3 =

16) 1 + 5 =

17) 4 + 9 =

18) 5 + 2 =

19) 3 + 5 =

20) 8 + 4 =

21) 6 + 3 =

22) 10 + 4 =

23) 2 + 5 =

24) 5 + 3 =

Day 6

PAGE A DAY MATH

Count ⇨ 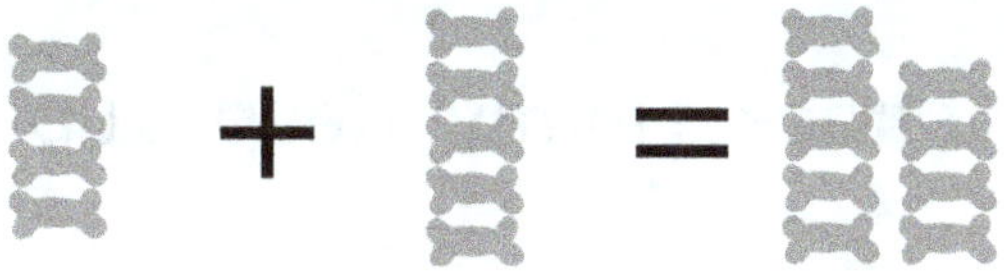

Learn ⇨ 4 + 5 = 9

Trace ⇨ 4 + 5 = 9

Copy ⇨ ☐ + ☐ = ☐

Bo says, "Try these...woof...go for it!"

1) 4 + 5 = ☐

2) 5 + 4 = ☐

3) 3 + 5 = ☐

4) 4 + 5 = ☐

5) 5 + 2 = ☐

6) 1 + 5 = ☐

7) 4 + 5 = ☐

8) 5 + 2 = ☐

9) 5 + 4 = ☐

10) 3 + 5 = ☐

Day 6

PAGE A DAY MATH

You are on the right track. Hurray! Keep it up!

11) 4 + 5 =

12) 2 + 5 =

13) 0 + 5 =

14) 5 + 4 =

15) 2 + 5 =

16) 5 + 1 =

17) 3 + 5 =

18) 5 + 2 =

19) 3 + 5 =

20) 5 + 4 =

21) 4 + 5 =

22) 1 + 5 =

23) 5 + 3 =

24) 4 + 5 =

Day 7

Count ⇨ 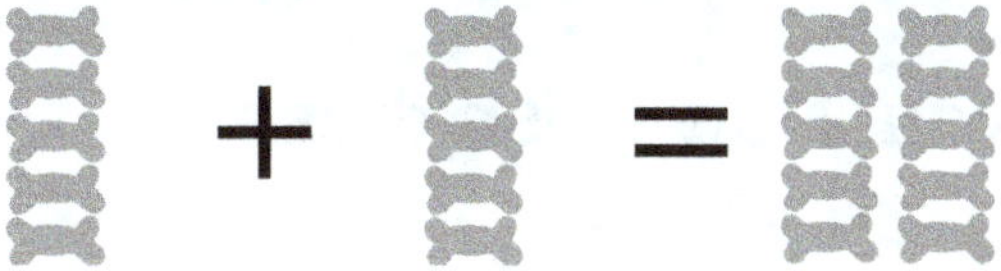

Learn ⇨ 5 + 5 = 10

Trace ⇨ 5 + 5 = 10

Copy ⇨ ☐ + ☐ = ☐

OK, now try these. Mo knows you can do it.

1) 5 + 5 = ☐

2) 5 + 4 = ☐

3) 5 + 5 = ☐

4) 4 + 5 = ☐

5) 5 + 5 = ☐

6) 5 + 3 = ☐

7) 5 + 5 = ☐

8) 3 + 5 = ☐

9) 2 + 5 = ☐

10) 5 + 5 = ☐

Day 7

PAGE A DAY MATH

Terrific! Good for you! Now finish these.

11) 5 + 5 = ☐

12) 1 + 5 = ☐

13) 3 + 5 = ☐

14) 8 + 4 = ☐

15) 5 + 5 = ☐

16) 2 + 5 = ☐

17) 4 + 5 = ☐

18) 5 + 2 = ☐

19) 5 + 5 = ☐

20) 1 + 5 = ☐

21) 5 + 3 = ☐

22) 9 + 4 = ☐

23) 5 + 5 = ☐

24) 5 + 4 = ☐

Day 8 Review

PAGE A DAY MATH

You are trying so hard. Great effort!

1) 9 + 2 = ☐

2) 4 + 7 = ☐

3) 3 + 4 = ☐

4) 6 + 2 = ☐

5) 4 + 8 = ☐

6) 3 + 3 = ☐

7) 9 + 4 = ☐

8) 3 + 8 = ☐

9) 6 + 3 = ☐

10) 3 + 5 = ☐

11) 7 + 3 = ☐

12) 2 + 2 = ☐

13) 8 + 3 = ☐

14) 4 + 4 = ☐

Day 8 Review

PAGE A DAY MATH

You are becoming a super hero math star. Awesome!

15) 9 + 2 =

16) 4 + 4 =

17) 3 + 9 =

18) 1 + 7 =

19) 4 + 2 =

20) 2 + 7 =

21) 5 + 3 =

22) 3 + 3 =

23) 3 + 4 =

24) 2 + 9 =

25) 6 + 4 =

26) 9 + 1 =

27) 4 + 5 =

28) 5 + 5 =

Count ⇨ 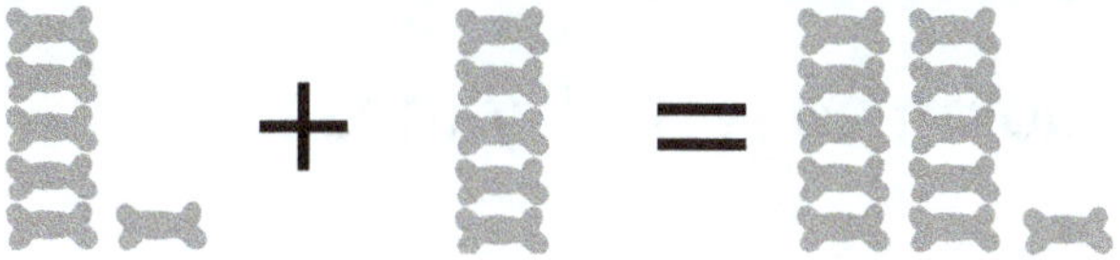

Learn ⇨ 6 + 5 = 11

Trace ⇨ 6 + 5 = 11

Copy ⇨ ☐ + ☐ = ☐

You are really improving. Bark-bark.

1) 6 + 5 = ☐

2) 5 + 6 = ☐

3) 6 + 4 = ☐

4) 6 + 5 = ☐

5) 5 + 6 = ☐

6) 4 + 6 = ☐

7) 6 + 5 = ☐

8) 5 + 5 = ☐

9) 5 + 6 = ☐

10) 4 + 5 = ☐

Day 9

PAGE A DAY MATH

Super! You are doing so well. Try these.

11) 6 + 5 =

12) 5 + 3 =

13) 1 + 5 =

14) 5 + 5 =

15) 5 + 2 =

16) 6 + 5 =

17) 5 + 4 =

18) 4 + 5 =

19) 0 + 5 =

20) 5 + 6 =

21) 3 + 5 =

22) 5 + 2 =

23) 1 + 5 =

24) 5 + 5 =

Day 10

PAGE A DAY MATH

Count ⇨ 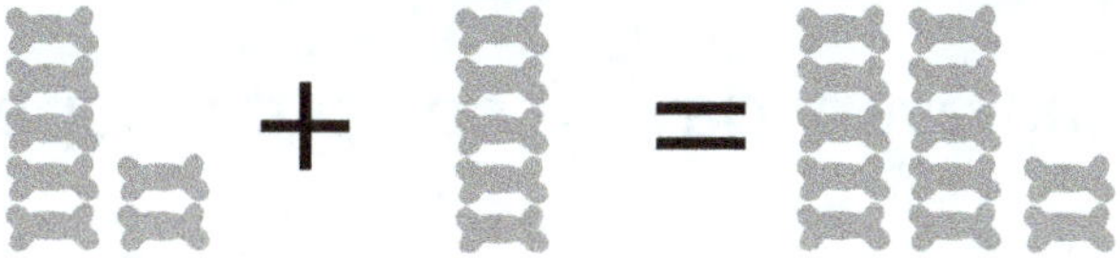

Learn ⇨ 7 + 5 = 12

Trace ⇨ 7 + 5 = 12

Copy ⇨ ☐ + ☐ = ☐

You are so determined. Good for you! Ruff!

1) 7 + 5 = ☐

2) 5 + 7 = ☐

3) 6 + 5 = ☐

4) 7 + 5 = ☐

5) 5 + 7 = ☐

6) 5 + 6 = ☐

7) 5 + 5 = ☐

8) 7 + 5 = ☐

9) 4 + 5 = ☐

10) 5 + 7 = ☐

Day 10

You are improving every day. Keep up the super effort.

11) 4 + 5 =

12) 7 + 5 =

13) 5 + 1 =

14) 5 + 6 =

15) 5 + 3 =

16) 2 + 5 =

17) 5 + 7 =

18) 1 + 5 =

19) 7 + 5 =

20) 2 + 5 =

21) 3 + 5 =

22) 4 + 5 =

23) 5 + 7 =

24) 6 + 5 =

Day 11

Count ⇨ 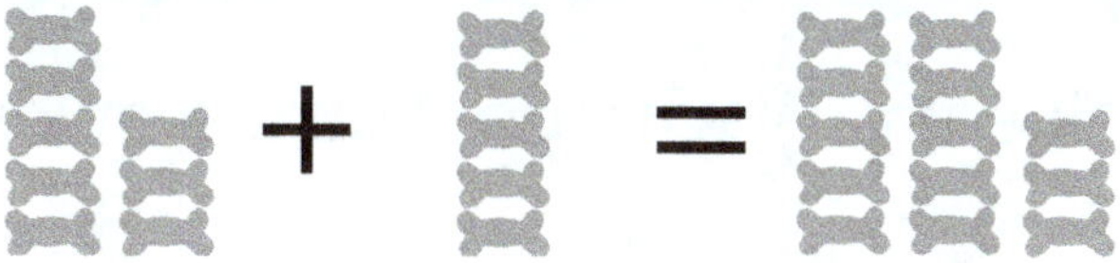

Learn ⇨ 8 + 5 = 13

Trace ⇨ 8 + 5 = 13

Copy ⇨ ☐ + ☐ = ☐

You have learned so much. Now try these.

1) 8 + 5 = ☐

2) 5 + 8 = ☐

3) 7 + 5 = ☐

4) 8 + 5 = ☐

5) 5 + 5 = ☐

6) 5 + 7 = ☐

7) 5 + 8 = ☐

8) 6 + 5 = ☐

9) 8 + 5 = ☐

10) 5 + 6 = ☐

You make it look easy. Way to go. You are awesome!

11) 8 + 5 =

12) 7 + 5 =

13) 5 + 3 =

14) 6 + 5 =

15) 5 + 8 =

16) 4 + 5 =

17) 5 + 7 =

18) 5 + 4 =

19) 1 + 5 =

20) 8 + 5 =

21) 5 + 0 =

22) 5 + 5 =

23) 2 + 5 =

24) 5 + 6 =

Day 12 Review

PAGE A DAY MATH

You are a math star! Tremendous. Keep it up!

1) 9 + 4 = ☐

2) 3 + 10 = ☐

3) 5 + 5 = ☐

4) 9 + 3 = ☐

5) 4 + 10 = ☐

6) 8 + 5 = ☐

7) 7 + 3 = ☐

8) 3 + 8 = ☐

9) 1 + 2 = ☐

10) 4 + 5 = ☐

11) 8 + 4 = ☐

12) 3 + 6 = ☐

13) 5 + 7 = ☐

14) 7 + 4 = ☐

Day 12 Review

You have it now. Keep up the effort. Go for it.

15) 5 + 3 =

16) 4 + 4 =

17) 2 + 5 =

18) 1 + 4 =

19) 7 + 2 =

20) 6 + 4 =

21) 2 + 3 =

22) 1 + 3 =

23) 4 + 5 =

24) 9 + 2 =

25) 2 + 4 =

26) 3 + 4 =

27) 3 + 3 =

28) 2 + 8 =

Day 13

Count ⇨ 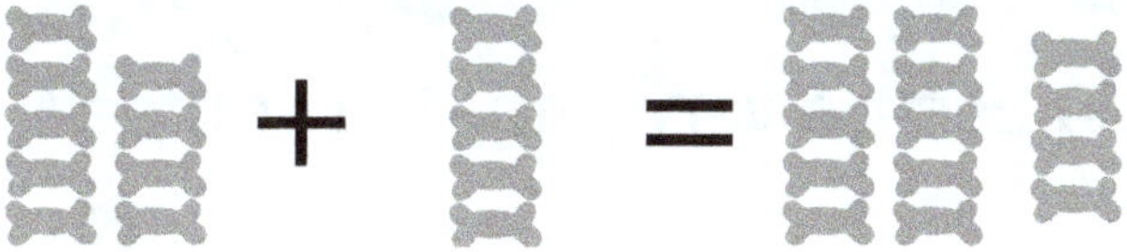

Learn ⇨ 9 + 5 = 14

Trace ⇨ 9 + 5 = 14

Copy ⇨ ☐ + ☐ = ☐

You have the hang of it. You're great at math!

1) 9 + 5 = ☐

2) 5 + 9 = ☐

3) 8 + 5 = ☐

4) 9 + 5 = ☐

5) 5 + 9 = ☐

6) 5 + 8 = ☐

7) 9 + 5 = ☐

8) 7 + 5 = ☐

9) 5 + 9 = ☐

10) 5 + 7 = ☐

Nice going. You can be very proud of yourself. Wow!

11) 5 + 7 =

12) 9 + 5 =

13) 5 + 6 =

14) 4 + 5 =

15) 5 + 9 =

16) 7 + 5 =

17) 8 + 5 =

18) 7 + 5 =

19) 5 + 4 =

20) 3 + 5 =

21) 5 + 5 =

22) 5 + 8 =

23) 6 + 5 =

24) 5 + 7 =

Day 14

PAGE A DAY MATH

Count ⇨ 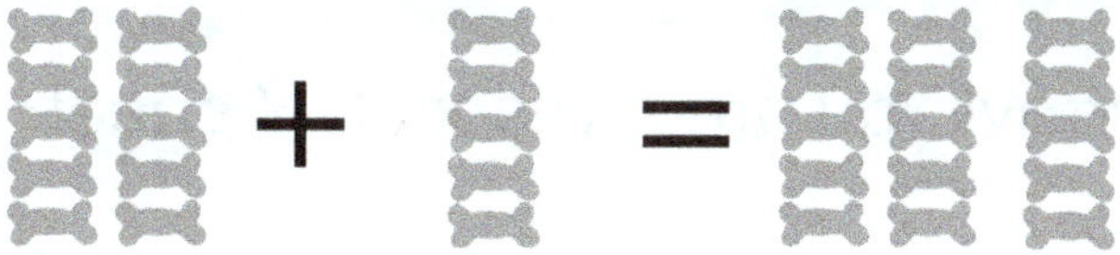

Learn ⇨ 10 + 5 = 15

Trace ⇨ 10 + 5 = 15

Copy ⇨ ___ + ___ = ___

You are almost finished with this book! Great!

1) 10 + 5 = ___

2) 5 + 10 = ___

3) 9 + 5 = ___

4) 10 + 5 = ___

5) 5 + 10 = ___

6) 5 + 9 = ___

7) 8 + 5 = ___

8) 6 + 5 = ___

9) 10 + 5 = ___

10) 5 + 10 = ___

Day 14

Last page! Hurray! You are awesome! Yippee!

11) 10 + 5 =

12) 5 + 7 =

13) 9 + 5 =

14) 5 + 6 =

15) 5 + 10 =

16) 5 + 5 =

17) 8 + 5 =

18) 5 + 9 =

19) 8 + 5 =

20) 10 + 5 =

21) 5 + 7 =

22) 9 + 5 =

23) 6 + 5 =

24) 5 + 10 =

HURRAY! YOU ARE A MATH STAR!

THE MATH SQUAD CONGRATULATES ____________________

FOR COMPLETING **ADDITION AND COUNTING, BOOK 5.**

www.ingramcontent.com/pod-product-compliance
Lightning Source LLC
Chambersburg PA
CBHW081403080526
44588CB00016B/2576